INTERNATIONAL UNION OF PURE AND APPLIED CHEMISTRY

MANUAL OF SYMBOLS AND TERMINOLOGY FOR PHYSICOCHEMICAL QUANTITIES AND UNITS

1969

INTERNATIONAL UNION OF
PURE AND APPLIED CHEMISTRY
DIVISION OF PHYSICAL CHEMISTRY
COMMISSION ON SYMBOLS, TERMINOLOGY,
AND UNITS

MANUAL OF SYMBOLS AND TERMINOLOGY FOR PHYSICOCHEMICAL QUANTITIES AND UNITS

Adopted by the IUPAC Council at Cortina d'Ampezzo, Italy, on 7 July 1969

Prepared for publication by

M. L. McGlashan
Chairman, Commission on Symbols, Terminology, and Units

LONDON
BUTTERWORTHS

ENGLAND: BUTTERWORTH & CO. (PUBLISHERS) LTD.
LONDON: 88 Kingsway, WC.2

AUSTRALIA: BUTTERWORTH & CO. (AUSTRALIA) LTD.
SYDNEY: 586 Pacific Highway Chatswood, NSW 2067
MELBOURNE: 343 Little Collins Street, 3000
BRISBANE: 240 Queen Street, 4000

CANADA: BUTTERWORTH & CO. (CANADA) LTD.
TORONTO: 14 Curity Avenue, 374

NEW ZEALAND: BUTTERWORTH & CO. (NEW ZEALAND) LTD.
WELLINGTON: 26/28 Waring Taylor Street, 1
AUCKLAND: 35 High Street, 1

SOUTH AFRICA: BUTTERWORTH & CO. (SOUTH AFRICA) LTD.
DURBAN: 152/154 Gale Street

The contents of this book appear in

Pure and Applied Chemistry, Vol. 21. No. 1 (1970)

Suggested U.D.C. *number:* 541. 1.081: 001.4+003.62

Second impression 1971

Standard Book Number 408 89350 8

Printed in Great Britain by Page Bros. (Norwich) Ltd., Norwich

PREFACE

The Commission on Symbols, Terminology, and Units† is a part of the Division of Physical Chemistry of the International Union of Pure and Applied Chemistry. Its general responsibilities are to secure clarity and precision, and wider agreement in the use of symbols, by chemists in different countries, among physicists, chemists, and engineers, and by editors of scientific journals. In pursuing these aims, liaison is maintained with other international organizations and in particular with the Commission on Symbols, Units and Nomenclature of the International Union of Pure and Applied Physics (SUN Commission) and Technical Committee 12 of the International Organization for Standardization (ISO/TC 12). References to the publications of these two organizations are given in 13.1 and 13.2 of this Manual. These publications may be referred to for more extended coverage of symbols for quantities, and related information, not commonly used by chemists. The recommendations presented here are generally in agreement with those of the SUN Commission and ISO/TC 12.

The present publication supersedes the Commission's publication of 1959 (Reference 13.3) in English and French and its translations into other languages.

M. L. McGlashan
Chairman
Commission on Symbols, Terminology, and Units

Department of Chemistry
University of Exeter
Exeter
August 1969

† The membership of the Commission during the period 1963 to 1969 in which the present Manual was prepared was as follows.

Titular Members
Chairman: 1963–1967 G. Waddington (U.S.A.); 1967–1969 M. L. McGlashan (U.K.).
Secretary: 1963–1967 H. Brusset (France); 1967–1969 M. A. Paul (U.S.A.).
Members: 1965–1969 K. V. Astachov (U.S.S.R.); 1963–1969 R. G. Bates (U.S.A.); 1963 F. Daniels (U.S.A.); 1967–1969 M. Fayard (France); 1963–1965 J. I. Gerassimov (U.S.S.R.); 1963–1969 W. Jaenicke (Germany); 1967–1969 F. Jellinek (Netherlands); 1967–1969 A. Pérez Masiá (Spain); 1963–1967 M. L. McGlashan (U.K.); 1963–1967 M. Milone (Italy); 1963–1967 K. J. Pedersen (Denmark); 1967–1969 L. G. Sillén (Sweden); 1963–1967 E. H. Wiebenga (Netherlands).

Associate Members
1963–1969 Syûzô Seki (Japan); 1967–1969 G. Waddington (U.S.A.).

CONTENTS

1. PHYSICAL QUANTITIES AND SYMBOLS FOR PHYSICAL QUANTITIES

1.1 Physical quantities

A *physical quantity* is the product of a *numerical value* (a pure number) and a *unit*.

1.2 Basic physical quantities

Seven physical quantities are generally regarded as independent basic physical quantities. These seven and the symbols used to denote them are as follows:

basic physical quantity	*symbol for quantity*
length	l
mass	m
time	t
electric current	I
thermodynamic temperature	T
luminous intensity	I_{v}
amount of substance	n

Luminous intensity is seldom if ever needed in physical chemistry.

One of these independent basic quantities is of special importance to chemists but until recently no name had been given to it, although units such as the mole have been used for it. The name "amount of substance" is now reserved for this quantity.

The definition of amount of substance, as of all other physical quantities (see Section 5), has nothing to do with any choice of *unit*, and in particular has nothing to do with the particular unit of amount of substance called the mole (see Section 3.6). It is now as inconsistent to call n the "number of moles" as it is to call m the "number of kilogrammes" or l the "number of metres", since n, m, and l are symbols for quantities not for numbers.

The amount of a substance is proportional to the number of *specified* elementary units of that substance. The proportionality factor is the same for all substances; its reciprocal is the Avogadro constant. The specified elementary unit may be an atom, a molecule, an ion, a radical, an electron, etc., or any *specified* group of such particles.

1.3 Derived physical quantities

All other physical quantities are regarded as being derived from, and as having dimensions derived from, the seven independent basic physical quantities by definitions involving only multiplication, division, differentiation, and/or integration. Examples of derived physical quantities are given, often with brief definitions, in Section 2.

1.4 Use of the words "specific" and "molar" in the names of physical quantities

The word "specific" before the name of an extensive physical quantity is restricted to the meaning "divided by mass". For example specific volume is the volume divided by the mass. When the extensive quantity is represented by a capital letter, the corresponding specific quantity may be represented by the corresponding lower case letter.

Examples: volume: V
specific volume: $v = V/m$
heat capacity at constant pressure: C_p
specific heat capacity at constant pressure: $c_p = C_p/m$

The word "molar" before the name of an extensive quantity is restricted to the meaning "divided by amount of substance". For example molar volume is the volume divided by the amount of substance. The subscript m attached to the symbol for the extensive quantity denotes the corresponding molar quantity.

Examples: volume: V — molar volume: $V_\mathrm{m} = V/n$
Gibbs energy: G — molar Gibbs energy: $G_\mathrm{m} = G/n$

The subscript m may be omitted when there is no risk of ambiguity. Lower case letters may be used to denote molar quantities when there is no risk of misinterpretation.

The symbol X_B, where X denotes an extensive quantity and B is the chemical symbol for a substance, denotes the partial molar quantity of the substance B defined by the relation:

$$X_\mathrm{B} = (\partial X/\partial n_\mathrm{B})_{T,p,n_\mathrm{C},\ldots}$$

For a pure substance B the partial molar quantity X_B and the molar quantity X_m are identical. The partial molar quantity X_B of pure substance B, which is identical with the molar quantity X_m of pure substance B, may be denoted by X_B^*, where the superscript * denotes "pure", so as to distinguish it from the partial molar quantity X_B of substance B in a mixture.

1.5 Printing of symbols for physical quantities

The symbols for physical quantities should be single letters[1] of the Latin or Greek alphabets which, when necessary, may be modified by subscripts

[1] An exception to this rule has been made for certain numbers used in the study of transport processes (see Section 2.9), for which the internationally agreed symbols consist of two letters. When such a symbol appears as a factor in a product, it is recommended that it be separated from the other symbols by a space, by a multiplication sign, or by parentheses. *Example:* Reynolds number: *Re*

and superscripts of specified meaning. The symbols for physical quantities should always be printed in italic (sloping) type.

The symbols for vector quantities should be printed in bold-faced italic type.

1.6 Printing of subscripts and superscripts

Subscripts or superscripts which are themselves symbols for physical quantities or numbers should be printed in italic (sloping) type and all others in roman (upright) type.

Example: C_p for heat capacity at constant pressure, but
C_{B} for heat capacity of substance B

1.7 Products and quotients of physical quantities

A product of two quantities a and b may be represented in any of the ways:

$$ab \quad \text{or} \quad a \cdot b \quad \text{or} \quad a\,.\,b \quad \text{or} \quad a \times b$$

and their quotient in any of the ways:

$$\frac{a}{b} \quad \text{or} \quad a/b \quad \text{or} \quad ab^{-1}$$

or in any of the other ways of writing the product of a and b^{-1}.

These rules may be extended to more complex groupings but more than one solidus (/) should never be used in the same expression unless parentheses are used to eliminate ambiguity.

Example: $(a/b)/c$ or $a/(b/c)$ but never $a/b/c$

2. RECOMMENDED NAMES AND SYMBOLS FOR QUANTITIES IN CHEMISTRY AND PHYSICS

The following list contains the recommended symbols for the most important quantities likely to be used by chemists. Whenever possible the symbol used for a physical quantity should be that recommended. In a few cases where conflicts were foreseen alternative recommendations have been made. Bold-faced italic (sloping) as well as ordinary italic (sloping) type can also sometimes be used to resolve conflicts. Further flexibility can be obtained by the use of capital letters as variants for lower-case letters, and *vice versa*, when no ambiguity is thereby introduced.

For example, d and D may be used instead of d_i and d_e for internal and external diameter in a context in which no quantity appears, such as diffusion coefficient, for which the recommended symbol is D. Again, the recommended symbol for power is P and for pressure is p or P, but P and p may be used for two powers or for two pressures; if power and pressure appear together, however, P should be used only for power and p only for pressure, and necessary distinctions between different powers or between different pressures should be made by the use of subscripts or other modifying signs.

When the above recommendations are insufficient to resolve a conflict or where a need arises for other reasons, an author is of course free to choose an *ad hoc* symbol. Any *ad hoc* symbol should be particularly carefully defined.

In the following list, where two or more symbols are indicated for a given quantity and are separated only by commas (without parentheses), they are on an equal footing; symbols within parentheses are reserve symbols.

Any description given after the name of a physical quantity is merely for identification and is not intended to be a complete definition.

Vector notation (bold-faced italic or sloping type) is used where appropriate in Section 2.6; it may be used when convenient also for appropriate quantities in other Sections.

2.1 Space, time, and related quantities

2.1.01	length	l
2.1.02	height	h
2.1.03	radius	r
2.1.04	diameter	d
2.1.05	path, length of arc	s
2.1.06	wavelength	λ
2.1.07	wavenumber: $1/\lambda$	σ, $\tilde{\nu}$ [1]
2.1.08	plane angle	$\alpha, \beta, \gamma, \theta, \phi$
2 1.09	solid angle	ω, Ω

(1) In solid-state studies k is used.

2.1.10	area	A, S, A_s (1)
2.1.11	volume	V
2.1.12	time	t
2.1.13	frequency	ν, f
2.1.14	angular frequency, pulsatance: $2\pi\nu$	ω
2.1.15	period: $1/\nu$	T
2.1.16	characteristic time interval, relaxation time, time constant	τ
2.1.17	velocity	v, u, w, c
2.1.18	angular velocity: $d\phi/dt$	ω
2.1.19	acceleration	a
2.1.20	acceleration of free fall	g

2.2 Mechanical and related quantities

2.2.01	mass	m
2.2.02	reduced mass	μ
2.2.03	specific volume (volume divided by mass)	v
2.2.04	density (mass divided by volume)	ρ
2.2.05	relative density (ratio of the density to that of a reference substance)	d
2.2.06	moment of inertia	I
2.2.07	momentum	p
2.2.08	force	F
2.2.09	weight	$G, (W)$
2.2.10	moment of force	M
2.2.11	angular momentum	L
2.2.12	work (force times path)	w, W
2.2.13	energy	E
2.2.14	potential energy	E_p, V, Φ
2.2.15	kinetic energy	E_k, T, K
2.2.16	Hamiltonian function	H
2.2.17	Lagrangian function	L
2.2.18	power (energy divided by time)	P
2.2.19	pressure	p, P
2.2.20	normal stress	σ
2.2.21	shear stress	τ
2.2.22	linear strain (relative elongation): $\Delta l/l_0$	ϵ, e
2.2.23	volume strain (bulk strain): $\Delta V/V_0$	θ
2.2.24	modulus of elasticity (normal stress divided by linear strain, Young's modulus)	E
2.2.25	shear modulus (shear stress divided by shear angle)	G
2.2.26	compressibility: $-V^{-1}(dV/dp)$	κ
2.2.27	compression (bulk) modulus: $(p = -K\Delta V/V_0)$	K
2.2.28	velocity of sound	c
2.2.29	viscosity	$\eta, (\mu)$

(1) The symbol A_s may be used when necessary to avoid confusion with the symbol A for Helmholtz energy.

2.2.30	fluidity: $1/\eta$	ϕ
2.2.31	kinematic viscosity: η/ρ	ν
2.2.32	friction coefficient (frictional force divided by normal force)	μ, (f)
2.2.33	surface tension	γ, σ
2.2.34	angle of contact	θ
2.2.35	diffusion coefficient	D
2.2.36	mass transfer coefficient (mass divided by time and by cross-sectional area)	k, k_m

2.3 Molecular and related quantities

2.3.01	relative atomic mass of an element (also called "atomic weight")[(1)]	A_r
2.3.02	relative molecular mass of a substance (also called "molecular weight")[(2)]	M_r
2.3.03	molar mass (mass divided by amount of substance)	M
2.3.04	Avogadro constant	L, N_A
2.3.05	number of molecules	N
2.3.06	amount of substance[(3)]	n, (ν)
2.3.07	mole fraction of substance B: $n_B/\Sigma_i n_i$	x_B, y_B
2.3.08	mass fraction of substance B	w_B
2.3.09	volume fraction of substance B	ϕ_B
2.3.10	molality of solute substance B (amount of B divided by mass of solvent)[(4)]	m_B
2.3.11	concentration of solute substance B (amount of B divided by the volume of the solution)[(5)]	c_B, [B]
2.3.12	mass concentration of substance B (mass of B divided by the volume of the solution)	ρ_B
2.3.13	surface concentration, surface excess	Γ
2.3.14	collision diameter of a molecule	d, σ
2.3.15	mean free path	l, λ
2.3.16	collision number (number of collisions divided by volume and by time)	Z

(1) The ratio of the average mass per atom of the natural nuclidic composition of an element to 1/12 of the mass of an atom of nuclide ^{12}C.
Example: $A_r(Cl) = 35.453$
The concept of relative atomic mass may be extended to other specified nuclidic compositions, but the natural nuclidic composition is assumed unless some other composition is specified.

(2) The ratio of the average mass per formula unit of the natural nuclidic composition of a substance to 1/12 of the mass of an atom of nuclide ^{12}C.
Example: $M_r(KCl) = 74.555$
The concept of relative molecular mass may be extended to other specified nuclidic compositions, but the natural nuclidic composition is assumed unless some other composition is specified.

(3) See Section 1.2.

(4) A solution having a molality equal to 0.1 mol kg^{-1} is sometimes called a 0.1 molal solution or a 0.1 m solution. If a symbol is needed for mol kg^{-1} then some symbol other than m, which is the SI unit-symbol for the metre, should be chosen.

(5) Concentration is sometimes called "molarity" but this name is both unnecessary and liable to cause confusion with molality and is therefore not recommended. A solution with a concentration of 0.1 mol dm^{-3} is often called a 0.1 molar solution or a 0.1 M solution.

2.3.17	grand partition function (system)	Ξ
2.3.18	partition function (system)	Q, Z
2.3.19	partition function (particle)	q, z
2.3.20	statistical weight	g
2.3.21	symmetry number	σ, s
2.3.22	characteristic temperature	Θ

2.4 Thermodynamic and related quantities

2.4.01	thermodynamic temperature, absolute temperature	T
2.4.02	Celsius temperature	t, θ
2.4.03	(molar) gas constant	R
2.4.04	Boltzmann constant	k
2.4.05	heat	q, Q (1)
2.4.06	work	w, W (1)
2.4.07	internal energy	U, (E)
2.4.08	enthalpy: $U + pV$	H
2.4.09	entropy	S
2.4.10	Helmholtz energy: $U - TS$	A
2.4.11	Massieu function: $-A/T$	J
2.4.12	Gibbs energy: $H - TS$	G
2.4.13	Planck function: $-G/T$	Y
2.4.14	compression factor: pV_m/RT	Z
2.4.15	heat capacity	C
2.4.16	specific heat capacity (heat capacity divided by mass; the name "specific heat" is not recommended)	c
2.4.17	ratio C_p/C_V	γ, (κ)
2.4.18	Joule-Thomson coefficient	μ
2.4.19	thermal conductivity	λ, k
2.4.20	thermal diffusivity: $\lambda/\rho c_p$	a
2.4.21	coefficient of heat transfer (density of heat flow rate divided by temperature difference)	h
2.4.22	cubic expansion coefficient: $V^{-1}(\partial V/\partial T)_p$	α
2.4.23	isothermal compressibility: $-V^{-1}(\partial V/\partial p)_T$	κ
2.4.24	pressure coefficient: $(\partial p/\partial T)_V$	β
2.4.25	chemical potential of substance B	μ_B
2.4.26	absolute activity of substance B: $\exp(\mu_B/RT)$	λ_B
2.4.27	fugacity	f, (p^*)
2.4.28	osmotic pressure	Π
2.4.29	ionic strength: ($I_m = \frac{1}{2}\Sigma_i m_i z_i^2$ or $I_c = \frac{1}{2}\Sigma_i c_i z_i^2$)	I
2.4.30	activity, relative activity of substance B	a_B
2.4.31	activity coefficient, mole fraction basis	f_B
2.4.32	activity coefficient, molality basis	γ_B
2.4.33	activity coefficient, concentration basis	y_B
2.4.34	osmotic coefficient	ϕ

(1) It is recommended that $q > 0$ and $w > 0$ both indicate *increase* of energy of the system under discussion. Thus $\Delta U = q + w$.

2.5 Chemical reactions

2.5.01	stoichiometric coefficient of substance B (negative for reactants, positive for products)	ν_B
2.5.02	general equation for a chemical reaction	$0 = \Sigma_B \nu_B B$
2.5.03	extent of reaction: $(d\xi = dn_B/\nu_B)$	ξ
2.5.04	rate of reaction: $d\xi/dt$ (see Section 11)	$\dot{\xi}$, J
2.5.05	rate of increase of concentration of substance B: dc_B/dt	v_B, r_B
2.5.06	rate constant	k
2.5.07	affinity of a reaction: $-\Sigma_B \nu_B \mu_B$	$\boldsymbol{A}$, $(\mathscr{A})$
2.5.08	equilibrium constant	K
2.5.09	degree of dissociation	α

2.6 Electricity and magnetism

2.6.01	elementary charge (of a proton)	e
2.6.02	quantity of electricity	Q
2.6.03	charge density	ρ
2.6.04	surface charge density	σ
2.6.05	electric current	I
2.6.06	electric current density	j
2.6.07	electric potential	V, ϕ
2.6.08	electric tension: IR	U
2.6.09	electric field strength	$\boldsymbol{E}$
2.6.10	electric displacement	$\boldsymbol{D}$
2.6.11	capacitance	C
2.6.12	permittivity: $(\boldsymbol{D} = \epsilon \boldsymbol{E})$	ϵ
2.6.13	permittivity of vacuum	ϵ_0
2.6.14	relative permittivity(1): ϵ/ϵ_0	ϵ_r, (ϵ)
2.6.15	dielectric polarization: $\boldsymbol{D} - \epsilon_0 \boldsymbol{E}$	$\boldsymbol{P}$
2.6.16	electric susceptibility: $\epsilon_r - 1$	χ_e
2.6.17	electric dipole moment	$\boldsymbol{p}$, $\boldsymbol{p}_e$
2.6.18	permanent dipole moment of a molecule	$\boldsymbol{p}$, $\boldsymbol{\mu}$
2.6.19	induced dipole moment of a molecule	$\boldsymbol{p}$, $\boldsymbol{p}_i$
2.6.20	electric polarizability of a molecule	α
2.6.21	magnetic flux	Φ
2.6.22	magnetic flux density, magnetic induction	$\boldsymbol{B}$
2.6.23	magnetic field strength	$\boldsymbol{H}$
2.6.24	permeability: $(\boldsymbol{B} = \mu \boldsymbol{H})$	μ
2.6.25	permeability of vacuum	μ_0
2.6.26	relative permeability: μ/μ_0	μ_r
2.6.27	magnetization: $(\boldsymbol{B}/\mu_0) - \boldsymbol{H}$	$\boldsymbol{M}$
2.6.28	magnetic susceptibility: $\mu_r - 1$	χ, (χ_m)
2.6.29	Bohr magneton	μ_B
2.6.30	electromagnetic moment: $(E_p = -\boldsymbol{m} \cdot \boldsymbol{B})$	$\boldsymbol{m}$, μ

(1) Also called dielectric constant, and sometimes denoted by D, when it is independent of $\boldsymbol{E}$.

2.6.31	resistance	R
2.6.32	resistivity (formerly called specific resistance): ($\boldsymbol{E} = \rho \boldsymbol{j}$)	ρ
2.6.33	conductivity (formerly called specific conductance): ($\boldsymbol{j} = \kappa \boldsymbol{E}$)	κ, (σ)
2.6.34	self-inductance	L
2.6.35	mutual inductance	M, L_{12}
2.6.36	reactance	X
2.6.37	impedance (complex impedance): $R + \mathrm{i}X$	Z
2.6.38	loss angle	δ
2.6.39	admittance (complex admittance): $1/Z$	Y
2.6.40	conductance: ($Y = G + \mathrm{i}B$)	G
2.6.41	susceptance: ($Y = G + \mathrm{i}B$)	B

2.7 Electrochemistry

2.7.01	Faraday constant	F
2.7.02	charge number of an ion B (positive for cations, negative for anions)	z_B
2.7.03	charge number of a cell reaction	z
2.7.04	electromotive force	E
2.7.05	electrochemical potential of ionic component B: $\mu_B + z_B F \phi$	$\tilde{\mu}_B$
2.7.06	electric mobility (velocity divided by electric field strength)	u, μ
2.7.07	electrolytic conductivity (formerly called specific conductance)	κ, (σ)
2.7.08	molar conductivity of electrolyte or ion(1): κ/c	Λ, λ (2)
2.7.09	transport number (transference number or migration number)	t
2.7.10	overpotential, overtension (also called "overvoltage")	η
2.7.11	exchange current density	j_0
2.7.12	electrochemical transfer coefficient	α
2.7.13	strength of double layer (electric moment divided by area)	τ
2.7.14	electrokinetic potential (zeta potential): τ/ϵ	ζ
2.7.15	thickness of diffusion layer	δ
2.7.16	inner electric potential	ϕ
2.7.17	outer electric potential	ψ
2.7.18	surface electric potential difference: $\phi - \psi$	χ

2.8 Light and related electromagnetic radiation

2.8.01	Planck constant	h
2.8.02	Planck constant divided by 2π	$\hbar$

(1) The word molar, contrary to the general rule given in Section 1.4, here means "divided by concentration".

(2) The formula unit whose concentration is c must be specified.
Example: $\Lambda(\mathrm{Mg}^{2+}) = 2\Lambda(\tfrac{1}{2}\mathrm{Mg}^{2+})$

2.8.03	radiant energy	Q [1]
2.8.04	radiant flux, radiant power	Φ [1]
2.8.05	radiant intensity: $\mathrm{d}\Phi/\mathrm{d}\omega$	I [1]
2.8.06	radiance: $(\mathrm{d}I/\mathrm{d}S)\cos\theta$	L [1]
2.8.07	radiant exitance: $\mathrm{d}\Phi/\mathrm{d}S$	M [1]
2.8.08	irradiance: $\mathrm{d}\Phi/\mathrm{d}S$	E [1]
2.8.09	absorptance, absorption factor[2] (ratio of absorbed to incident radiant or luminous flux)	α
2.8.10	reflectance, reflection factor[2] (ratio of reflected to incident radiant or luminous flux)	ρ
2.8.11	transmittance, transmission factor[2] (ratio of transmitted to incident radiant or luminous flux)	τ
2.8.12	internal transmittance[2] (transmittance of the medium itself, disregarding boundary or container influence)	τ_i, T
2.8.13	internal transmission density, absorbance[2]: $\log_{10}(1/\tau_\mathrm{i})$	D_i, A
2.8.14	(linear) absorption coefficient[2]: D_i/l	a
2.8.15	molar (linear) absorption coefficient[2, 3]: D_i/lc	ϵ
2.8.16	quantum yield	Φ
2.8.17	exposure: $\int E\mathrm{d}t$	H
2.8.18	speed of light in vacuo	c
2.8.19	refractive index	n
2.8.20	molar refraction: $(n^2 - 1)V_\mathrm{m}/(n^2 + 2)$	R_m
2.8.21	angle of optical rotation	α

2.9 Transport properties,[4]

2.9.01	flux (of a quantity X)	J_X, J
2.9.02	Reynolds number: $\rho vl/\eta$	Re
2.9.03	Euler number: $p/\rho v^2$	Eu
2.9.04	Froude number: $v/(lg)^{\frac{1}{2}}$	Fr
2.9.05	Grashof number: $l^3 g\alpha\Delta\theta\rho^2/\eta^2$	Gr
2.9.06	Weber number: $\rho v^2 l/\gamma$	We

(1) The same symbol is often used also for the corresponding luminous quantity. Subscripts e for energetic and v for visible may be added whenever confusion between these quantities might otherwise occur.

(2) These names and symbols are in agreement with those adopted jointly by the International Commission of Illumination (CIE) and the International Electrotechnical Commission (IEC). The terms extinction (for 2.8.13), extinction coefficient (for 2.8.14), and molar extinction coefficient (for 2.8.15) are unsuitable because extinction is now reserved for diffusion of radiation rather than absorption. The terms absorptivity (for 2.8.14) and molar absorptivity (for 2.8.15) should be avoided because the meaning absorptance per unit length has been accepted internationally for the term absorptivity.

(3) See Footnote (1) to Section 2.7.

(4) References to the symbols used in defining the quantities 2.9.02 to 2.9.26 are as follows:

ρ	2.2.04	v	2.1.17	l	2.1.01	η	2.2.29	p	2.2.19
g	2.1.20	α	2.4.22	θ	2.4.02	γ	2.2.33	c	2.2.28
λ	2.3.15	f	2.1.13	a	2.4.30	t	2.1.12	h	2.4.21
k	2.4.19	c_p	2.4.16	D	2.2.35	x	2.3.07	k_m	2.2.36
μ	2.6.24	κ	2.6.33	$\boldsymbol{B}$	2.6.22				

2.9.07	Mach number: v/c	*Ma*
2.9.08	Knudsen number: λ/l	*Kn*
2.9.09	Strouhal number: lf/v	*Sr*
2.9.10	Fourier number: $a\Delta t/l^2$	*Fo*
2.9.11	Peclet number: vl/a	*Pe*
2.9.12	Rayleigh number: $l^3 g\alpha\Delta\theta\rho/\eta a$	*Ra*
2.9.13	Nusselt number: hl/k	*Nu*
2.9.14	Stanton number: $h/\rho v c_p$	*St*
2.9.15	Fourier number for mass transfer: Dt/l^2	*Fo**
2.9.16	Peclet number for mass transfer: vl/D	*Pe**
2.9.17	Grashof number for mass transfer: $-l^3 g(\partial\rho/\partial x)_{T,p}\Delta x\rho/\eta^2$	*Gr**
2.9.18	Nusselt number for mass transfer: $k_m l/\rho D$	*Nu**
2.9.19	Stanton number for mass transfer: $k_m/\rho v$	*St**
2.9.20	Prandtl number: $\eta/\rho a$	*Pr*
2.9.21	Schmidt number: $\eta/\rho D$	*Sc*
2.9.22	Lewis number: a/D	*Le*
2.9.23	Magnetic Reynolds number: $v\mu\kappa l$	Re_{m}
2.9.24	Alfvén number: $vl(\rho\mu)^{\frac{1}{2}}/\boldsymbol{B}$	*Al*
2.9.25	Hartmann number: $\boldsymbol{B}l(\kappa/\eta)^{\frac{1}{2}}$	*Ha*
2.9.26	Cowling number: $\boldsymbol{B}^2/\mu\rho v^2$	*Co*

2.10 Symbols for particular cases of physical quantities

It is much more difficult to make detailed recommendations on symbols for physical quantities in particular cases than in general cases. The reason is the incompatibility between the need for specifying numerous details and the need for keeping the printing reasonably simple. Among the most awkward things to print are superscripts to subscripts and subscripts to subscripts. Examples of symbols to be avoided are:

$$\Lambda_{\mathrm{NO}_3^-} \qquad \Delta H_{25\,^\circ\mathrm{C}} \qquad (pV)^{p=0}_{0\,^\circ\mathrm{C}}$$

The problem is vastly reduced if it is recognized that two different kinds of notation are required for two different purposes. In the formulation of general fundamental relations the most important requirement is a notation which is easy to understand and easy to remember. In applications to particular cases, in quoting numerical values, and in tabulation, the most important requirement is complete elimination of any possible ambiguity even at the cost of an elaborate notation.

The advantage of a dual notation is already to some extent accepted in the case of concentration. The recommended notation for the formulation of the equilibrium constant K_c for the general reaction:

$$0 = \Sigma_{\mathrm{B}}\nu_{\mathrm{B}}\mathrm{B}$$

is

$$K_c = \Pi_{\mathrm{B}}(c_{\mathrm{B}})^{\nu_{\mathrm{B}}}$$

but when we turn to a particular example it is better to use a notation such as:

$$Br_2 + H_2O = HOBr + H^+ + Br^-$$

$$\frac{[HOBr]\,[H^+]\,[Br^-]}{[Br_2]} = K_c$$

$$K_c(25\ °C) = 6 \times 10^{-9}\ mol^2\ dm^{-6}$$

Once the principle of dual notation is accepted, its adaptability and usefulness become manifest in all fields of physical chemistry. It will here be illustrated by just a few examples.

The general relation between the molar conductivity of an electrolyte and the molar conductivities of the two ions is written most simply and most clearly as:

$$\Lambda = \Lambda^+ + \Lambda^-$$

but when it comes to giving values in particular cases a much more appropriate notation is:

$$\Lambda(\tfrac{1}{2}Mg^{2+}) = 53\ \Omega^{-1}\ cm^2\ mol^{-1}\ \text{at}\ 25\ °C$$
$$\Lambda(Cl^-) = 76\ \Omega^{-1}\ cm^2\ mol^{-1}\ \text{at}\ 25\ °C$$
$$\Lambda(\tfrac{1}{2}MgCl_2) = 129\ \Omega^{-1}\ cm^2\ mol^{-1}\ \text{at}\ 25\ °C$$
$$\Lambda(MgCl_2) = 258\ \Omega^{-1}\ cm^2\ mol^{-1}\ \text{at}\ 25\ °C$$

The general relation between the partial molar volumes of the two components A and B of a binary mixture is written most simply:

$$n_A dV_A + n_B dV_B = 0 \qquad (T, p\ \text{const.})$$

But when it comes to specifying values, a completely different notation is called for, such as:

$$V(K_2SO_4,\ 0.1\ mol\ dm^{-3}\ \text{in}\ H_2O,\ 25\ °C) = 48\ cm^3\ mol^{-1}$$

Each kind of notation is appropriate to its purpose.

A last example will be given relating to optical rotation. The relations between the angle α of rotation of the plane of polarization and the amount n, or the number N of molecules, of the optically active substance in the path of a light beam of cross-section A can be clearly expressed in the form:

$$\alpha = n\alpha_n/A = N\alpha_N/A$$

where α_n is the molar optical rotatory power and α_N the molecular optical rotatory power. When on the other hand it is desired to record an experimental measurement, an appropriate notation would be:

$$\alpha(589.3\ nm,\ 20\ °C,\ \text{sucrose},\ 10\ g\ dm^{-3}\ \text{in}\ H_2O,\ 10\ cm) = +66.470°$$

2.11 Recommended superscripts

The following superscripts are recommended.

- ° or * pure substance
- ∞ infinite dilution
- id ideal
- ° or ⊖ standard in general
- ‡ activated complex, transition state

3. UNITS AND SYMBOLS FOR UNITS

3.1 Printing of symbols for units

The symbol for a unit should be printed in roman (upright) type, should remain unaltered in the plural, and should not be followed by a full stop except when it occurs at the end of a sentence in text.

Example: 5 cm but not 5 cms and not 5 cm. and not 5 cms.

The symbol for a unit derived from a proper name should begin with a capital roman (upright) letter.

Examples: J for joule and Hz for hertz

Any other symbol for a unit should be printed in lower case roman (upright) type.

3.2 Printing of prefixes

Symbols for prefixes for units should be printed in roman (upright) type with no space between the prefix and the unit. Compound prefixes should be avoided. (See Section 3.11.)

Example: ns but not mμs for 10^{-9} s

3.3 Combination of prefixes and symbols

A combination of prefix and symbol for a unit is regarded as a single symbol which may be raised to a power without the use of brackets.

Examples: cm^2 means $(\mathrm{cm})^2$ and $\mu\mathrm{s}^{-1}$ means $(\mu\mathrm{s})^{-1}$

3.4 Multiplication and division of units

A product of two units may be represented in any of the ways:

$$\mathrm{N\,m} \quad \text{or} \quad \mathrm{N \cdot m} \quad \text{or} \quad \mathrm{N\,.\,m} \quad \text{or} \quad \mathrm{N \times m}$$

The representation Nm is not recommended.

A quotient of two units may be represented in any of the ways:

$$\frac{\mathrm{m}}{\mathrm{s}} \quad \text{or} \quad \mathrm{m/s} \quad \text{or} \quad \mathrm{m\,s^{-1}}$$

or in any of the other ways of writing the product of m and s^{-1}.

These rules may be extended to more complex groupings but more than one solidus (/) should never be used in the same expression unless parentheses are used to eliminate ambiguity.

Example: $\mathrm{J\,K^{-1}\,mol^{-1}}$ or J/(K mol) but never J/K/mol

3.5 The International System of Units

The name International System of Units has been adopted by the Conférence Générale des Poids et Mesures for the system of units based on a selected set of independent *SI Base Units.*

The SI Base Units are the metre, kilogramme, second, ampere, kelvin, candela, and mole.[1, 2] In the International System of Units there is one and only one *SI Unit* for each physical quantity. This is either the appropriate SI Base Unit itself (see Section 3.7) or the appropriate *SI Derived Unit* formed by multiplication and/or division of two or more SI Base Units (see Section 3.10). A few such SI Derived Units have been given special names and symbols (see Section 3.9). There are also two *SI Supplementary Units* for which it is not decided whether they are SI Base Units or SI Derived Units (see Section 3.8).

Any of the approved decimal prefixes, called *SI Prefixes*, may be used to construct decimal multiples or submultiples of SI Units (see Section 3.11).

It is recommended that only units composed of SI Units and SI Prefixes be used in science and technology.

3.6 Definitions of the SI Base Units

metre: The metre is the length equal to 1650 763.73 wavelengths in vacuum of the radiation corresponding to the transition between the levels $2p_{10}$ and $5d_5$ of the krypton-86 atom.

kilogramme: The kilogramme is the unit of mass; it is equal to the mass of the international prototype of the kilogramme.

second: The second is the duration of 9 192 631 770 periods of the radiation corresponding to the transition between the two hyperfine levels of the ground state of the caesium-133 atom.

ampere: The ampere is that constant current which, if maintained in two straight parallel conductors of infinite length, of negligible cross-section, and placed 1 metre apart in vacuum, would produce between these conductors a force equal to 2×10^{-7} newton per metre of length.

kelvin: The kelvin, unit of thermodynamic temperature, is the fraction 1/273.16 of the thermodynamic temperature of the triple point of water.(1)

candela: The candela is the luminous intensity, in the perpendicular direction, of a surface of 1/600 000 square metre of a black body at the temperature of freezing platinum under a pressure of 101 325 newtons per square metre.

mole: The mole is the amount of substance of a system which contains as many elementary entities as there are carbon atoms in 0.012 kilogramme of carbon-12.

Note: The elementary entities must be specified and may be atoms, molecules, ions, electrons, other particles, or specified groups of such particles.(2)

Examples: 1 mole of HgCl has a mass of 236.04 grammes
1 mole of Hg_2Cl_2 has a mass of 472.08 grammes
1 mole of Hg has a mass of 200.59 grammes
1 mole of Hg_2^{2+} has a mass of 401.18 grammes
1 mole of $\frac{1}{2}Ca^{2+}$ has a mass of 20.04 grammes
1 mole of CuZn has a mass of 128.92 grammes
1 mole of $Cu_{0.5}Zn_{0.5}$ has a mass of 64.46 grammes

(1) In October 1967 the thirteenth Conférence Générale des Poids et Mesures recommended that the kelvin, symbol K, be used both for thermodynamic temperature and for thermodynamic temperature interval, and that the unit-symbols °K and deg be abandoned.

(2) In October 1969 the Comité International des Poids et Mesures decided to propose to the Fourteenth Conférence Générale des Poids et Mesures (1971) that the mole be introduced as a Base Unit in the International System of Units with the definition given above.

1 mole of $Fe_{0.91}S$ has a mass of 82.88 grammes
1 mole of e^- has a mass of 5.4860×10^{-4} gramme
1 mole of a mixture containing 78.09 moles per cent of N_2, 20.95 moles per cent of O_2, 0.93 mole per cent of Ar, and 0.03 mole per cent of CO_2, has a mass of 28.964 grammes

3.7 Names and symbols for SI Base Units

physical quantity	*name of SI Unit*	*symbol for SI Unit*
length	metre	m
mass	kilogramme	kg
time	second	s
electric current	ampere	A
thermodynamic temperature	kelvin	K
luminous intensity	candela	cd
amount of substance(1)	mole	mol

3.8 Names and symbols for SI Supplementary Units

physical quantity	*name of SI Unit*	*symbol for SI Unit*
plane angle	radian	rad
solid angle	steradian	sr

3.9 Special names and symbols for certain SI Derived Units

physical quantity	*name of SI Unit*	*symbol for SI Unit*	*definition of SI Unit*
force	newton	N	$kg\ m\ s^{-2}$
pressure	pascal	Pa	$kg\ m^{-1}\ s^{-2}$ ($= N\ m^{-2}$) (2)
energy	joule	J	$kg\ m^2\ s^{-2}$
power	watt	W	$kg\ m^2\ s^{-3}$ ($= J\ s^{-1}$)
electric charge	coulomb	C	A s
electric potential difference	volt	V	$kg\ m^2\ s^{-3}\ A^{-1}$ ($= J\ A^{-1}\ s^{-1}$)
electric resistance	ohm	Ω	$kg\ m^2\ s^{-3}\ A^{-2}$ ($= V\ A^{-1}$)
electric conductance	siemens	S	$kg^{-1}\ m^{-2}\ s^3\ A^2$ ($= A\ V^{-1} = \Omega^{-1}$) (2)
electric capacitance	farad	F	$A^2\ s^4\ kg^{-1}\ m^{-2}$ ($= A\ s\ V^{-1}$)
magnetic flux	weber	Wb	$kg\ m^2\ s^{-2}\ A^{-1}$ ($= V\ s$)
inductance	henry	H	$kg\ m^2\ s^{-2}\ A^{-2}$ ($= V\ A^{-1}\ s$)
magnetic flux density	tesla	T	$kg\ s^{-2}\ A^{-1}$ ($= V\ s\ m^{-2}$)
luminous flux	lumen	lm	cd sr
illumination	lux	lx	$cd\ sr\ m^{-2}$
frequency	hertz	Hz	s^{-1}

(1) See Section 1.2.

(2) In June 1969 the Comité Consultatif des Unités recommended that these names and these symbols be approved for these SI Units.

3.10 SI Derived Units and Unit-symbols for other quantities

(This list is not exhaustive.)

physical quantity	*SI Unit*	*symbol for SI Unit*
area	square metre	m^2
volume	cubic metre	m^3
density	kilogramme per cubic metre	$kg\ m^{-3}$
velocity	metre per second	$m\ s^{-1}$
angular velocity	radian per second	$rad\ s^{-1}$
acceleration	metre per second squared	$m\ s^{-2}$
pressure	newton per square metre	$N\ m^{-2}$
kinematic viscosity, diffusion coefficient	square metre per second	$m^2\ s^{-1}$
dynamic viscosity	newton-second per square metre	$N\ s\ m^{-2}$
molar entropy, molar heat capacity	joule per kelvin mole	$J\ K^{-1}\ mol^{-1}$
concentration	mole per cubic metre	$mol\ m^{-3}$
electric field strength	volt per metre	$V\ m^{-1}$
magnetic field strength	ampere per metre	$A\ m^{-1}$
luminance	candela per square metre	$cd\ m^{-2}$

3.11 SI Prefixes

fraction	*prefix*	*symbol*	*multiple*	*prefix*	*symbol*
10^{-1}	deci	d	10	deca	da
10^{-2}	centi	c	10^2	hecto	h
10^{-3}	milli	m	10^3	kilo	k
10^{-6}	micro	μ	10^6	mega	M
10^{-9}	nano	n	10^9	giga	G
10^{-12}	pico	p	10^{12}	tera	T
10^{-15}	femto	f			
10^{-18}	atto	a			

The names and symbols of decimal multiples and sub-multiples of the unit of mass, which already contains a prefix, are constructed by adding the appropriate prefix to the word gramme and symbol g:

Examples: mg not μkg   μg not nkg   Mg not kkg

3.12 The degree Celsius

physical quantity	*name of unit*	*symbol for unit*	*definition of unit*
Celsius temperature	degree Celsius	°C (1)	$t/°C = T/K - 273.15$

(1) The ° sign and the letter following form one symbol and there should be no space between them.
Example: 25 °C not 25° C.

3.13 Decimal fractions and multiples of SI Units having special names

These units do not belong to the International System of Units. Their use is to be progressively discouraged. The time-scale implied by the word "progressively" need not, however, be the same for all these units, nor for any unit need it be the same in all fields of science. In the meantime it is recommended that any author who uses any of these units will define them in terms of SI Units once in each publication in which he uses them. The following list is not exhaustive.

physical quantity	*name of unit*	*symbol for unit*	*definition of unit*
length	ångström	Å	10^{-10} m
length	micron[(1)]	μ	10^{-6} m = μm
area	barn	b	10^{-28} m^2
volume	litre[(2)]	l	10^{-3} m^3
mass	tonne	t	10^3 kg
force	dyne	dyn	10^{-5} N
pressure	bar	bar	10^5 N m^{-2}
energy	erg	erg	10^{-7} J
kinematic viscosity, diffusion coefficient	stokes	St	10^{-4} m^2 s^{-1}
dynamic viscosity	poise	P	10^{-1} kg m^{-1} s^{-1}
concentration	mole per litre	M	10^3 mol m^{-3}
magnetic flux	maxwell	Mx	10^{-8} Wb
magnetic flux density (magnetic induction)	gauss	G	10^{-4} T

3.14 Some other units now exactly defined in terms of the SI Units

These units do not belong to the International System of Units and their use is to be progressively discouraged and eventually abandoned. In the meantime it is recommended that any author who uses any of these units will define them in terms of SI Units once in each publication in which he uses them. Each of the definitions given in the fourth column is *exact.*

physical quantity	*name of unit*	*symbol for unit*	*definition of unit*
length	inch	in	2.54×10^{-2} m
mass	pound (avoirdupois)	lb	0.453 592 37 kg
force	kilogramme-force	kgf	9.806 65 N
pressure	atmosphere	atm	101 325 N m^{-2}
pressure	torr	Torr	(101 325/760) N m^{-2}

(1) The name micron, symbol μ, is still used instead of its SI equivalent the micrometre, symbol μm, and likewise the millimicron, symbol mμ, instead of the nanometre, symbol nm.

(2) By decision of the twelfth Conférence Générale des Poids et Mesures in October 1964 the old definition of the litre (1.000 028 dm^3) was rescinded. The word litre is now regarded as a special name for the cubic decimetre. Neither the word litre nor its symbol l should be used to express results of high precision.

pressure	conventional millimetre of mercury[1]	mmHg	$13.5951 \times 980.665 \times 10^{-2}$ N m^{-2}
energy	kilowatt-hour	kW h	3.6×10^{6} J
energy	thermochemical calorie	cal_{th}	4.184 J
energy	I.T. calorie	cal_{IT}	4.1868 J
thermodynamic temperature	degree Rankine	°R	(5/9) K
radioactivity	curie	Ci	3.7×10^{10} s^{-1}

3.15 Units defined in terms of the best available experimental values of certain physical constants

These units do not belong to the International System of Units and their use is to be progressively discouraged. The factors for conversion of these units to SI Units are subject to change in the light of new experimental measurements of the constants involved. The following list is not exhaustive.

physical quantity	*name of unit*	*symbol for unit*	*conversion factor*
energy	electronvolt	eV	eV $\approx 1.6021 \times 10^{-19}$ J
mass	(unified) atomic mass unit	u	u $\approx 1.66041 \times 10^{-27}$ kg

3.16 "International" electrical units

These units are obsolete having been replaced by the "absolute" (SI) units in 1948. The conversion factors which should be used with electrical measurements quoted in "international" units depend on where and on when the instruments used to make the measurements were calibrated. The following two sets of conversion factors refer respectively to the "mean international" units estimated by the ninth Conférence Générale des Poids et Mesures in 1948, and to the "U.S. international" units estimated by the National Bureau of Standards (U.S.A.) as applying to published measurements made with instruments calibrated by them prior to 1948.

1 "mean international ohm" = 1.00049 Ω
1 "mean international volt" = 1.00034 V
1 "U.S. international ohm" = 1.000495 Ω
1 "U.S. international volt" = 1.000330 V

3.17 Electrical and magnetic units belonging to unit-systems other than the International System of Units

Definitions of units used in the obsolescent "electrostatic CGS" and "electromagnetic CGS" unit-systems can be found in References 13.1.05 and 13.2.

(1) The conventional millimetre of mercury, symbol mmHg (not mm Hg), is the pressure exerted by a column exactly 1 mm high of a fluid of density exactly 13.5951 g cm^{-3} in a place where the gravitational acceleration is exactly 980.665 cm s^{-2}. The mmHg differs from the Torr by less than 2×10^{-7} Torr.

4. NUMBERS

4.1 Printing of numbers

Numbers should be printed in upright type. The decimal sign between digits in a number should be a comma (,) or (but *only* in English-language texts) a point(.). To facilitate the reading of long numbers the digits may be grouped in threes but no comma or point should ever be used except for the decimal sign.

Example: 2 573,421 736 or in English language texts 2 573.421 736 but never 2,573.421,736

When the decimal sign is placed before the first digit of a number a zero should always be placed before the decimal sign.

Example: $0{,}2573 \times 10^4$ or in English language texts 0.2573×10^4 but not $,2573 \times 10^4$ and not $.2573 \times 10^4$

It is often convenient to print numbers with just one digit before the decimal sign.

Example: $2{,}573 \times 10^3$ or in English-language text 2.573×10^3

4.2 Multiplication and division of numbers

The multiplication sign between numbers should be a cross (×) or (but never when a point is used as the decimal sign) a centred dot (·).

Example: 2.3×3.4 or $2{,}3 \cdot 3{,}4$

Division of one number by another may be indicated in any of the ways:

$$\frac{136}{273} \quad \text{or} \quad 136/273 \quad \text{or} \quad 136 \times (273)^{-1}$$

These rules may be extended to more complex groupings, but more than one solidus (/) should never be used in the same expression unless parentheses are used to eliminate ambiguity.

Example: $(136/273)/2.303$ or $136/(273 \times 2.303)$ but never $136/273/2.303$

5. PHYSICAL QUANTITIES, UNITS, AND NUMERICAL VALUES

As stated in Section 1.1 the value of a *physical quantity* is equal to the product of a *numerical value* and a *unit*:

$$\text{physical quantity} = \text{numerical value} \times \text{unit}.$$

Neither any physical quantity, nor the symbol used to denote it, should imply a particular choice of unit.

Operations on equations involving physical quantities, units, and numerical values, should follow the ordinary rules of algebra.

Thus the physical quantity called the critical pressure and denoted by p_c has the value for water:

$$p_c = 221.2 \text{ bar} \quad \text{or better} \quad p_c = 22.12 \text{ MPa}.$$

These equations may equally well be written in the forms:

$$p_c/\text{bar} = 221.2 \quad \text{or better} \quad p_c/\text{MPa} = 22.12,$$

which are especially useful for the headings in tables and as labels on the axes of graphs.

6. RECOMMENDED MATHEMATICAL SYMBOLS[1]

Mathematical operators (for example d and Δ) and mathematical constants (for example e and π) should always be printed in roman (upright) type. Letter symbols for numbers other than mathematical constants should be printed in italic type.

equal to	$=$
not equal to	$\neq$ $\neq$
identically equal to	$\equiv$
corresponds to	$\triangleq$
approximately equal to	$\approx$
approaches	$\rightarrow$
asymptotically equal to	$\simeq$
proportional to	$\propto$ $\sim$
infinity	∞
less than	$<$
greater than	$>$
less than or equal to	$\leqq$ $\leqslant$ $\lesssim$
greater than or equal to	$\geqq$ $\geqslant$ $\gtrsim$
much less than	$\ll$
much greater than	$\gg$
plus	$+$
minus	$-$
multiplied by	$\times$ $\cdot$
a divided by *b*	$\frac{a}{b}$ a/b ab^{-1}
magnitude of *a*	$\lvert a \rvert$
a raised to the power *n*	a^n
square root of *a*	$a^{1/2}$ $a^{\frac{1}{2}}$ $\sqrt{a}$ $\sqrt{a}$
n'th root of *a*	$a^{1/n}$ $a^{\frac{1}{n}}$ $\sqrt[n]{a}$ $\sqrt[n]{a}$
mean value of *a*	$\langle a \rangle$ $\bar{a}$
natural logarithm of *a*	$\ln a$ $\log_{e} a$
decadic logarithm of *a*	$\lg a$ $\log_{10} a$ $\log a$
binary logarithm of *a*	$\mathrm{lb}\, a$ $\log_2 a$
exponential of *a*	$\exp a$ e^a

(1) Taken from Reference 13.1.11 where a more comprehensive list can be found.

7. SYMBOLS FOR CHEMICAL ELEMENTS, NUCLIDES, AND PARTICLES

7.1 Definitions

A nuclide is a species of atoms of which each atom has identical atomic number (proton number) and identical mass number (nucleon number). Different nuclides having the same value of the atomic number are named isotopes or isotopic nuclides. Different nuclides having the same mass number are named isobars or isobaric nuclides.

7.2 Elements and nuclides

Symbols for chemical elements should be written in roman (upright) type. The symbol is not followed by a full stop except when it occurs at the end of a sentence in text.

Examples: Ca C H He

The nuclide may be specified by attaching numbers. The mass number should be placed in the left superscript position; the atomic number, if desired, may be placed as a left subscript. The number of atoms per molecule is indicated as a right subscript. Ionic charge, or state of excitation, or oxidation number(1) may be indicated in the right superscript space.

Examples: Mass number: $^{14}N_2$, $^{35}Cl^-$
Ionic charge: Cl^-, Ca^{2+}, PO_4^{3-} or $PO_4{}^{3-}$
Excited electronic state: He*, NO*
Oxidation number: $Pb_2^{II}Pb^{IV}O_4$, $K_6M^{IV}Mo_9O_{32}$ (where M denotes a metal)

7.3 Particles

neutron	n	α-particle	α
proton	p	electron	e
deuteron	d	photon	γ
triton	t		

The electric charge of particles may be indicated by adding the superscript +, —, or 0; e.g., p^+, n^0, e^+, e^-. If the symbols p and e are used without charge, they refer to the positive proton and negative electron respectively.

7.4 Abbreviated notation for nuclear reactions

The meaning of the symbolic expression indicating a nuclear reaction should be the following:

$$\begin{matrix}\text{initial} \\ \text{nuclide}\end{matrix}\left(\begin{matrix}\text{incoming particle(s)} \\ \text{or quanta}\end{matrix}, \begin{matrix}\text{outgoing particle(s)} \\ \text{or quanta}\end{matrix}\right)\begin{matrix}\text{final} \\ \text{nuclide}\end{matrix}$$

Examples: $^{14}N(\alpha,p)^{17}O$ $^{59}Co(n,\gamma)^{60}Co$
$^{23}Na(\gamma,3n)^{20}Na$ $^{31}P(\gamma,pn)^{29}Si$

(1) For a more detailed discussion see Reference 13.4.

8. SYMBOLS FOR SPECTROSCOPY[1]

8.1 General rules

A letter-symbol indicating the quantum state of *a system* should be printed in capital upright type. A letter-symbol indicating the quantum state of *a single particle* should be printed in lower case upright type.

8.2 Atomic spectroscopy

The letter-symbols indicating quantum states are:

$L, l = 0$: S, s	$L, l = 4$: G, g	$L, l = 8$: L, l
$= 1$: P, p	$= 5$: H, h	$= 9$: M, m
$= 2$: D, d	$= 6$: I, i	$= 10$: N, n
$= 3$: F, f	$= 7$: K, k	$= 11$: O, o

A right-hand subscript indicates the total angular momentum quantum number J or j. A left-hand superscript indicates the spin multiplicity $2S + 1$.

Examples: $^2\mathrm{P}_{3/2}$ — state ($J = 3/2$, multiplicity 2)
$\mathrm{p}_{3/2}$ — electron ($j = 3/2$)

An atomic electron configuration is indicated symbolically by:

$$(nl)^{\kappa}(n'l')^{\kappa'} \ldots$$

Instead of $l = 0, 1, 2, 3, \ldots$ one uses the quantum state symbols s, p, d, f, . . .

Example: the atomic configuration: $(1\mathrm{s})^2(2\mathrm{s})^2(2\mathrm{p})^3$

8.3 Molecular spectroscopy

The letter symbols indicating molecular electronic quantum states are in the case of *linear molecules:*

$$\Lambda, \lambda = 0: \Sigma, \sigma$$
$$= 1: \Pi, \pi$$
$$= 2: \Delta, \delta$$

and for *non-linear molecules:*

A, a; B, b; E, e; etc.

Remarks: A left-hand superscript indicates the spin multiplicity. For molecules having a symmetry centre the parity symbol g or u, indicating respectively symmetric or antisymmetric behaviour on inversion, is attached as a right-hand subscript. A + or − sign attached as a right-hand superscript indicates the symmetry as regards reflection in any plane through the symmetry axis of the molecules.

[1] Taken from Reference 13.2.

Examples: Σ_g^+, Π_u, $^2\Sigma$, $^3\Pi$, etc.

The letter symbols indicating the vibrational angular momentum states in the case of *linear molecules* are:

$$l = 0: \Sigma$$
$$= 1: \Pi$$
$$= 2: \Delta$$

8.4 Spectroscopic transitions

The upper level and the lower level are indicated by ′ and ″ respectively.

Example: $h\nu = E' - E''$

A spectroscopic transition should be indicated by writing the upper state first and the lower state second, connected by a dash in between.

Examples: $^2P_{1/2}$—$^2S_{1/2}$ for an electronic transition
$(J',K')-(J'',K'')$ for a rotational transition
$v'-v''$ for a vibrational transition

Absorption transition and emission transition may be indicated by arrows ← and → respectively.

Examples: $^2P_{1/2} \rightarrow {}^2S_{1/2}$ emission from $^2P_{1/2}$ to $^2S_{1/2}$
$(J',K') \leftarrow (J'',K'')$ absorption from (J'',K'') to (J',K')

The difference Δ between two quantum numbers should be that of the upper state minus that of the lower state.

Example: $\Delta J = J' - J''$

The indications of the branches of the rotation band should be as follows:

$$\Delta J = J' - J'' = -2: \text{O-branch}$$
$$= -1: \text{P-branch}$$
$$= 0: \text{Q-branch}$$
$$= +1: \text{R-branch}$$
$$= +2: \text{S-branch}$$

9. CONVENTIONS CONCERNING THE SIGNS OF ELECTRIC POTENTIAL DIFFERENCES, ELECTROMOTIVE FORCES, AND ELECTRODE POTENTIALS[1]

9.1 The electric potential difference for a galvanic cell

The cell should be represented by a diagram, for example:

$$Zn \mid Zn^{2+} \mid Cu^{2+} \mid Cu$$

The electric potential difference ΔV is equal in sign and magnitude to the electric potential of a metallic conducting lead on the right minus that of an identical lead on the left.

When the reaction of the cell is written as:

$$\tfrac{1}{2}Zn + \tfrac{1}{2}Cu^{2+} \rightarrow \tfrac{1}{2}Zn^{2+} + \tfrac{1}{2}Cu$$

this implies a diagram so drawn that this reaction takes place when positive electricity flows through the cell from left to right. If this is the direction of the current when the cell is short-circuited, as it will be in the present example (unless the ratio $[Cu^{2+}]/[Zn^{2+}]$ is extremely small), the electric potential difference will be positive.

If, however, the reaction is written as:

$$\tfrac{1}{2}Cu + \tfrac{1}{2}Zn^{2+} \rightarrow \tfrac{1}{2}Cu^{2+} + \tfrac{1}{2}Zn$$

this implies the diagram

$$Cu \mid Cu^{2+} \mid Zn^{2+} \mid Zn$$

and the electric potential difference of the cell so specified will be negative (unless the ratio $[Cu^{2+}]/[Zn^{2+}]$ is extremely small).

The limiting value of the electric potential difference for zero current through the cell is called the electromotive force and denoted by E.

9.2 Electrode potential

The so-called electrode potential of an electrode (half-cell) is defined as the electromotive force of a cell in which the electrode on the left is a *standard hydrogen electrode* and that on the right is the electrode in question. For example, for the zinc electrode (written as $Zn^{2+} \mid Zn$) the cell in question is:

$$Pt, H_2 \mid H^+ \mid Zn^{2+} \mid Zn$$

The reaction taking place at the zinc electrode is:

$$Zn^{2+} + 2e^- \rightarrow Zn$$

[1] The conventions given here are in accordance with the "Stockholm Convention" of 1953.

The latter is to be regarded as an abbreviation for the reaction in the mentioned cell:

$$Zn^{2+} + H_2 \rightarrow Zn + 2H^+$$

In the standard state the electromotive force of this cell has a negative sign and a value of − 0.763 V. The standard electrode potential of the zinc electrode is therefore − 0.763 V.

The symbol $Zn \mid Zn^{2+}$ on the other hand implies the cell:

$$Zn \mid Zn^{2+} \mid H^+ \mid H_2, Pt$$

in which the reaction is:

$$Zn + 2H^+ \rightarrow Zn^{2+} + H_2$$

The electromotive force of this cell should *not* be called an electrode potential.

10. THE QUANTITY pH [1]

10.1 Operational definition

In all existing national standards the definition of pH is an operational one. The electromotive force E_X of the cell:

Pt, H_2 | solution X | concentrated KCl solution | reference electrode

is measured and likewise the electromotive force E_S of the cell:

Pt, H_2 | solution S | concentrated KCl solution | reference electrode

both cells being at the same temperature throughout and the reference electrodes and bridge solutions being identical in the two cells. The pH of the solution X, denoted by pH(X), is then related to the pH of the solution S, denoted by pH(S), by the definition:

$$\mathrm{pH(X)} = \mathrm{pH(S)} + \frac{E_X - E_S}{(RT \ln 10)/F}$$

where R denotes the gas constant, T the thermodynamic temperature, and F the Faraday constant. Thus defined the quantity pH is a number.

To a good approximation, the hydrogen electrodes in both cells may be replaced by other hydrogen-ion-responsive electrodes, e.g. glass or quinhydrone. The two bridge solutions may be of any molality not less than 3.5 mol kg^{-1}, provided they are the same (see Reference 13.5).

10.2 Standards

The difference between the pH of two solutions having been defined as above, the definition of pH can be completed by assigning a value of pH at each temperature to one or more chosen solutions designated as standards. A series of pH(S) values for five suitable standard reference solutions is given in Section 10.3.

If the definition of pH given above is adhered to strictly, then the pH of a solution may be slightly dependent on which standard solution is used. These unavoidable deviations are caused not only by imperfections in the response of the hydrogen-ion electrodes but also by variations in the liquid junctions resulting from the different ionic compositions and mobilities of the several standards and from differences in the geometry of the liquid-liquid boundary. In fact such variations in measured pH are usually too small to be of practical significance. Moreover, the acceptance of several standards allows the use of the following alternative definition of pH.

The electromotive force E_X is measured, and likewise the electromotive forces E_1 and E_2 of two similar cells with the solution X replaced by the standard solutions S_1 and S_2 such that the E_1 and E_2 values are on either side of, and as near as possible to, E_X. The pH of solution X is then obtained

[1] The symbol pH is an exception to the general rules given in Section 1.5.

by assuming linearity between pH and E, that is to say

$$\frac{\mathrm{pH(X)} - \mathrm{pH(S_1)}}{\mathrm{pH(S_2)} - \mathrm{pH(S_1)}} = \frac{E_X - E_1}{E_2 - E_1}$$

This procedure is especially recommended when the hydrogen-ion-responsive electrode is a glass electrode.

10.3 Values of pH(S) for five standard solutions

t/°C	A	B	C	D	E
0		4.003	6.984	7.534	9.464
5		3.999	6.951	7.500	9.395
10		3.998	6.923	7.472	9.332
15		3.999	6.900	7.448	9.276
20		4.002	6.881	7.429	9.225
25	3.557	4.008	6.865	7.413	9.180
30	3.552	4.015	6.853	7.400	9.139
35	3.549	4.024	6.844	7.389	9.102
38	3.548	4.030	6.840	7.384	9.081
40	3.547	4.035	6.838	7.380	9.068
45	3.547	4.047	6.834	7.373	9.038
50	3.549	4.060	6.833	7.367	9.011
55	3.554	4.075	6.834		8.985
60	3.560	4.091	6.836		8.962
70	3.580	4.126	6.845		8.921
80	3.609	4.164	6.859		8.885
90	3.650	4.205	6.877		8.850
95	3.674	4.227	6.886		8.833

The compositions of the standard solutions are:

A: KH tartrate (saturated at 25 °C)
B: KH phthalate, $m = 0.05\ \mathrm{mol\ kg^{-1}}$
C: KH_2PO_4, $m = 0.025\ \mathrm{mol\ kg^{-1}}$;
Na_2HPO_4, $m = 0.025\ \mathrm{mol\ kg^{-1}}$
D: KH_2PO_4, $m = 0.008\,695\ \mathrm{mol\ kg^{-1}}$;
Na_2HPO_4, $m = 0.030\,43\ \mathrm{mol\ kg^{-1}}$
E: $Na_2B_4O_7$, $m = 0.01\ \mathrm{mol\ kg^{-1}}$

where m denotes molality and the solvent is water.

11. DEFINITION OF RATE OF REACTION AND RELATED QUANTITIES

11.1 Rate of reaction

For the reaction

$$0 = \Sigma_B \nu_B B$$

the extent of reaction ξ is defined according to 2.5.03 by

$$d\xi = \nu_B^{-1} dn_B$$

where n_B is the amount, and ν_B is the stoichiometric number, of the substance B.

It is recommended that the *rate of reaction* be defined as the rate of increase of the extent of reaction, namely

$$\dot{\xi} = d\xi/dt = \nu_B^{-1} dn_B/dt$$

This definition is independent of the choice of B and is valid regardless of the conditions under which a reaction is carried out, e.g. it is valid for a reaction in which the volume varies with time, or for a reaction involving two or more phases, or for a reaction carried out in a flow reactor.

If both sides of this equation are divided by any specified volume V, not necessarily independent of time, and not necessarily that of a single phase in which the reaction is taking place, then

$$V^{-1} d\xi/dt = V^{-1} \nu_B^{-1} dn_B/dt$$

If the specified volume V is independent of time, then

$$V^{-1} d\xi/dt = \nu_B^{-1} d(n_B/V)/dt$$

If this specified volume V is such that

$$n_B/V = c_B \text{ or } [B]$$

where c_B or [B] is the concentration of B, then

$$V^{-1} d\xi/dt = \nu_B^{-1} dc_B/dt \text{ or } \nu_B^{-1} d[B]/dt$$

The quantity

$$dn_B/dt \ (= \nu_B d\xi/dt)$$

may be called the rate of formation of B, and the quantity

$$V^{-1} \nu_B^{-1} dn_B/dt \ (= V^{-1} d\xi/dt)$$

may be called the rate of reaction divided by volume, and the quantity

$$v_B = dc_B/dt \text{ or } d[B]/dt$$

which has often been called the rate of reaction, may be called the rate of increase of the concentration of B, but none of these three quantities should be called the rate of reaction.

11.2 Order of reaction

If it is found *experimentally* that the rate of increase of the concentration of B is given by

$$v_B \propto [C]^c[D]^d \ldots$$

then the reaction is described as of order c with respect to C, of order d with respect to D, . . ., and of overall order $(c + d + \ldots)$.

11.3 Labelling of elementary processes

Elementary processes should be labelled in such a manner that reverse processes are immediately recognizable.

Example:

elementary process	*label*	*rate constant*
$Br_2 + M \rightarrow 2Br + M$	1	k_1
$Br + H_2 \rightarrow HBr + H$	2	k_2
$H + Br_2 \rightarrow HBr + Br$	3	k_3
$H + HBr \rightarrow H_2 + Br$	−2	k_{-2}
$2Br + M \rightarrow Br_2 + M$	−1	k_{-1}

11.4 Collision number

The collision number defined as the number of collisions per unit time and per unit volume and having dimensions $(\text{time})^{-1} \times (\text{volume})^{-1}$ should be denoted by Z.

The collision number divided by the product of two relevant concentrations (or by the square of the relevant concentration) and by the Avogadro constant is a second-order rate constant having dimensions $(\text{time})^{-1} \times (\text{volume}) \times (\text{amount of substance})^{-1}$ and should be denoted by z. Thus $z = Z/Lc_Ac_B$.

12. VALUES OF THE FUNDAMENTAL CONSTANTS

At the XXIIIrd Conference of IUPAC the Council recommended the use by chemists of a consistent set of fundamental constants which has been published (see Reference 13.6). For the details concerning the development of this set of constants and their assigned uncertainties the reader is referred to the original paper and references therein. For convenience the recommended values are given here for the most important constants. In each case the estimated uncertainty is three times the standard deviation.

quantity	*symbol*	*value (with estimated uncertainty)*
speed of light in vacuum	c	$2.997\,925 \times 10^{8}$ m s^{-1} $\pm 0.000\,003 \times 10^{8}$ m s^{-1}
Avogadro constant	L, N_A	$6.022\,52 \times 10^{23}$ mol^{-1} $\pm 0.000\,28 \times 10^{23}$ mol^{-1}
Faraday constant	F	$9.648\,70 \times 10^{4}$ C mol^{-1} $\pm 0.000\,16 \times 10^{4}$ C mol^{-1}
Planck constant	h	$6.625\,6 \times 10^{-34}$ J s $\pm 0.000\,5 \times 10^{-34}$ J s
"ice-point" temperature[1]	T_{ice}	273.150 0 K $\pm$ 0.000 1 K
	RT_{ice}	$2.271\,06 \times 10^{3}$ J mol^{-1} $\pm 0.000\,12 \times 10^{3}$ J mol^{-1}
gas constant	R	8.314 33 J K^{-1} mol^{-1} $\pm$0.000 44 J K^{-1} mol^{-1}
charge of proton	e	$1.602\,10 \times 10^{-19}$ C $\pm 0.000\,07 \times 10^{-19}$ C
Boltzmann constant	k	$1.380\,54 \times 10^{-23}$ J K^{-1} $\pm 0.000\,09 \times 10^{-23}$ J K^{-1}
second radiation constant	$c_2 = hc/k$	$1.438\,79 \times 10^{-2}$ m K $\pm 0.000\,09 \times 10^{-2}$ m K
Einstein constant relating mass and energy	$Y = c^2$	$8.987\,554 \times 10^{16}$ J kg^{-1} $\pm 0.000\,018 \times 10^{16}$ J kg^{-1}
constant relating wavenumber and energy	$Z = N_A hc$	$1.196\,255 \times 10^{-1}$ J m mol^{-1} $\pm 0.000\,038 \times 10^{-1}$ J m mol^{-1}
Bohr magneton	μ_B	$9.273\,2 \times 10^{-24}$ m^2 A $\pm 0.000\,6 \times 10^{-24}$ m^2 A
permeability of vacuum	μ_0	$4\pi \times 10^{-7}$ J s^2 C^{-2} m^{-1} (exactly)
permittivity of vacuum	$\epsilon_0 = \mu_0^{-1}c^{-2}$	$8.854\,185 \times 10^{-12}$ J^{-1} C^2 m^{-1} $\pm 0.000\,018 \times 10^{-12}$ J^{-1} C^2 m^{-1}

(1) The "ice-point" temperature T_{ice} is the temperature of equilibrium of solid and liquid water saturated with air at a pressure of one atmosphere. The quantity RT_{ice} is identical to the quantity $\lim_{p \to 0} (pV_m)$ for a gas at that temperature.

13. REFERENCES

13.1 ISO Recommendation R 31 will when complete form a comprehensive publication dealing with quantities and units in various fields of science and technology. The following parts have so far been published[1] and can be purchased in any country belonging to ISO from the "Member Body", usually the national standardizing organization of the country.

13.1.01 "Part I: Basic quantities and units of the SI", 2nd edition, December 1965.

13.1.02 "Part II: Quantities and units of periodic and related phenomena", 1st edition, February 1958.

13.1.03 "Part III: Quantities and units of mechanics", 1st edition, December 1960.

13.1.04 "Part IV: Quantities and units of heat", 1st edition, December 1960.

13.1.05 "Part V: Quantities and units of electricity and magnetism", 1st edition, November 1965.

13.1.07 "Part VII: Quantities and units of acoustics", 1st edition, November 1965.

13.1.11 "Part XI: Mathematical signs and symbols for use in physical sciences and technology", 1st edition, February 1961.

13.2 "Symbols, Units and Nomenclature in Physics", Document UIP 11 (SUN 65-3), published by IUPAP, 1965. This document supersedes Document UIP 9 (SUN 61-44) with the same title, which was published by IUPAP in 1961.

13.3 "Manual of Physicochemical Symbols and Terminology", published for IUPAC by Butterworths Scientific Publications, London, 1959. This document was reprinted in the *Journal of the American Chemical Society* (1960) **82**, 5517.

13.4 "Nomenclature of Inorganic Chemistry", published for IUPAC by Butterworths Scientific Publications, London, 1959. This document was reprinted in the *Journal of the American Chemical Society* (1960) **82**, 5523.

13.5 *Pure and Applied Chemistry* (1960), **1**, 163.

13.6 *Pure and Applied Chemistry* (1964), **9**, 453.

(1) Parts not yet published are Part 0: General principles concerning quantities, units, and symbols: Part VI: Quantities and units of light and related electromagnetic radiation; Part VIII: Quantities and units of physical chemistry and molecular physics; Part IX Quantities and units of atomic and nuclear physics; Part X: Quantities and units of nuclear reactions and ionizing radiations; Part XII: Dimensionless parameters.

APPENDIX I

DEFINITION OF ACTIVITIES AND RELATED QUANTITIES

A.I.1 *Chemical potential and absolute activity*

The chemical potential μ_B of a substance B in a mixture of substances B, C, . . ., is defined by

$$\mu_B = (\partial G/\partial n_B)_{T,p,n_C,\ \cdots}$$

where G is the Gibbs energy of the mixture, T is the thermodynamic temperature, p is the pressure, and n_B, n_C, . . ., are the amounts of the substances B, C, . . ., in the mixture.

(In molecular theory the symbol μ_B is sometimes used for the quantity μ_B/L where L is the Avogadro constant, but this usage is not recommended.)

The absolute activity λ_B of the substance B in the mixture is a number defined by

$$\lambda_B = \exp(\mu_B/RT) \quad \text{or} \quad \mu_B = RT \ln \lambda_B$$

where R is the gas constant.

The definitions given below often take simpler, though perhaps less familiar, forms when they are expressed in terms of absolute activity rather than in terms of chemical potential. Each of the definitions given below is expressed in both of these ways.

1. Pure substances

A.I.2 *Properties of pure substances*

The superscript * attached to the symbol for a property of a substance denotes the property of the *pure* substance. It is sometimes convenient to treat a mixture of constant composition as a pure substance.

A.I.3 *Fugacity of a pure gaseous substance*

The fugacity f_B^* of a pure gaseous substance B is a quantity with the same dimensions as pressure, defined in terms of the absolute activity λ_B^* of the pure gaseous substance B by

$$f_B^* = \lambda_B^* \lim_{p\to 0}(p/\lambda_B^*) \qquad (T \text{ const.})$$

or in terms of the chemical potential μ_B by

$$RT \ln f_B^* = \mu_B^* + \lim_{p\to 0}(RT \ln p - \mu_B^*) \qquad (T \text{ const.})$$

where p is the pressure of the gas and T is its thermodynamic temperature. It follows from this definition that

$$\lim_{p\to 0}(f_B^*/p) = 1 \qquad (T \text{ const.})$$

and that

$$RT \ln(f^*_B/p) = \int_0^p (V^*_B - RT/p)\, dp \qquad (T \text{ const.})$$

where V^*_B is the molar volume of the pure gaseous substance B.

A pure gaseous substance B is treated as an *ideal gas* when the approximation $f^*_B = p$ is used. The ratio (f^*_B/p) may be called the fugacity coefficient. The name activity coefficient has sometimes been used for this ratio but is not recommended.

2. Mixtures

A.I.4 *Definition of a mixture*

The word *mixture* is used to describe a gaseous or liquid or solid phase containing more than one substance, when the substances are all treated in the same way (contrast the use of the word *solution* in Section A.I.9).

A.I.5 *Partial pressure*

The partial pressure p_B of a substance B in a *gaseous* mixture is a quantity with the same dimensions as pressure defined by

$$p_B = y_B p$$

where y_B is the mole fraction of the substance B in the gaseous mixture and p is the pressure.

A.I.6 *Fugacity of a substance in a gaseous mixture*

The fugacity f_B of the substance B in a gaseous mixture containing mole fractions y_B, y_C, . . ., of the substances B, C, . . ., is a quantity with the same dimensions as pressure, defined in terms of the absolute activity λ_B of the substance B in the gaseous mixture by

$$f_B = \lambda_B \lim_{p \to 0} (y_B p/\lambda_B) \qquad (T \text{ const.})$$

or in terms of the chemical potential μ_B by

$$RT \ln f_B = \mu_B + \lim_{p \to 0} \{RT \ln(y_B p) - \mu_B\} \qquad (T \text{ const.})$$

It follows from this definition that

$$\lim_{p \to 0} (f_B/y_B p) = 1 \qquad (T \text{ const.})$$

and that

$$RT \ln(f_B/y_B p) = \int_0^p (V_B - RT/p)\, dp \qquad (T \text{ const.})$$

where V_B is the partial molar volume (see Section 1.4) of the substance B in the gaseous mixture.

A gaseous mixture of B, C, . . ., is treated as an *ideal gaseous mixture* when the approximations $f_B = y_B p$, $f_C = y_C p$, . . ., are used. It follows that $pV = (n_B + n_C + \ldots)RT$ for an ideal gaseous mixture of B, C, . . .

The ratio $(f_B/y_B p)$ may be called the fugacity coefficient of the substance B. The name activity coefficient has sometimes been used for this ratio but is not recommended.

When $y_B = 1$ the definitions given in this Section for the fugacity of a substance in a gaseous mixture reduce to those given in Section A.I.3 for the fugacity of a pure gaseous substance.

A.I.7 *Activity coefficient of a substance in a liquid or solid mixture*

The activity coefficient f_B of a substance B in a liquid or solid mixture containing mole fractions x_B, x_C, . . ., of the substances B, C, . . ., is a number defined in terms of the absolute activity λ_B of the substance B in the mixture by

$$f_B = \lambda_B / \lambda_B^* x_B$$

where λ_B^* is the absolute activity of the pure substance B at the same temperature and pressure, or in terms of the chemical potential μ_B by

$$RT \ln(x_B f_B) = \mu_B - \mu_B^*$$

where μ_B^* is the chemical potential of the pure substance B at the same temperature and pressure.

It follows from this definition that

$$\lim_{x_B \to 1} f_B = 1 \qquad (T, p \text{ const.})$$

A.I.8 *Relative activity of a substance in a liquid or solid mixture*

The relative activity a_B of a substance B in a liquid or solid mixture is a number defined by

$$a_B = \lambda_B / \lambda_B^*$$

or by

$$RT \ln a_B = \mu_B - \mu_B^*$$

where the other symbols are as defined in Section A.I.7.

It follows from this definition that

$$\lim_{x_B \to 1} a_B = 1 \qquad (T, p \text{ const.})$$

A mixture of substances B, C, . . ., is treated as an *ideal mixture* when the approximations $a_B = x_B$, $a_C = x_C$, . . ., and consequently $f_B = 1$, $f_C = 1$, . . ., are used.

3. Solutions

A.I.9 *Definition of a solution*

The word *solution* is used to describe a liquid or solid phase containing more than one substance, when for convenience one of the substances, which is called the *solvent* and may itself be a mixture, is treated differently from the other substances, which are called *solutes*. When, as is often but not necessarily the case, the sum of the mole fractions of the solutes is small compared with unity, the solution is called a *dilute solution*. In the following definitions the solvent substance is denoted by A and the solute substances by B, C, ...

A.I.10 *Properties of infinitely dilute solutions*

The superscript $^\infty$ attached to the symbol for a property of a solution denotes the property of an *infinitely dilute solution.*

For example if V_B denotes the partial molar volume (see Section 1.4) of the solute substance B in a solution containing molalities m_B, m_C, . . ., or mole fractions x_B, x_C, . . ., of solute substances B, C, . . ., in a solvent substance A, then

$$V_B^\infty = \lim_{\Sigma_i m_i \to 0} V_B = \lim_{\Sigma_i x_i \to 0} V_B \qquad (T, p \text{ const.})$$

where i = B, C, . . .
Similarly if V_A denotes the partial molar volume of the *solvent* substance A, then

$$V_A^\infty = \lim_{\Sigma_i m_i \to 0} V_A = \lim_{\Sigma_i x_i \to 0} V_A = V_A^* \qquad (T, p \text{ const.})$$

where V_A^* is the molar volume of the pure solvent substance A.

A.I.11 *Activity coefficient of a solute substance in a solution*

The activity coefficient γ_B of a *solute* substance B in a solution (especially in a dilute liquid solution) containing molalities m_B, m_C, . . ., of solute substances B, C, . . ., in a solvent substance A, is a number defined in terms of the absolute activity λ_B of the solute substance B in the solution by

$$\gamma_B = (\lambda_B/m_B)/(\lambda_B/m_B)^\infty \qquad (T, p \text{ const.})$$

or in terms of the chemical potential μ_B by

$$RT \ln(m_B\gamma_B) = \mu_B - (\mu_B - RT \ln m_B)^\infty \qquad (T, p \text{ const.})$$

It follows from this definition that

$$\gamma_B^\infty = 1 \qquad (T, p \text{ const.})$$

The name activity coefficient with the symbol y_B may be used for the quantity similarly defined but with concentration c_B (see Section 2.3) in place of molality m_B.

Another activity coefficient, called the *rational activity coefficient* of a solute substance B and denoted by $f_{x,B}$ is sometimes used. It is defined in terms of the absolute activity λ_B by

$$f_{x,B} = (\lambda_B/x_B)/(\lambda_B/x_B)^\infty \qquad (T, p \text{ const.})$$

or in terms of the chemical potential μ_B by

$$RT \ln(x_B f_{x,B}) = \mu_B - (\mu_B - RT \ln x_B)^\infty \qquad (T, p \text{ const.})$$

where x_B is the mole fraction of the solute substance B in the solution. The rational activity coefficient $f_{x,B}$ is related to the (practical) activity coefficient γ_B by the formula:

$$f_{x,B} = \gamma_B(1 + M_A\Sigma_i m_i) = \gamma_B/(1 - \Sigma_i x_i)$$

A solution of solute substances B, C, . . . , in a solvent substance A is treated as an *ideal dilute solution* when the activity coefficients are approximated to unity, for example $\gamma_B = 1$, $\gamma_C = 1$, ...

A.I.12 *Relative activity of a solute substance in a solution*

The relative activity a_B of a *solute* substance B in a solution (especially in a dilute liquid solution) containing molalities m_B, m_C, . . ., of solute substances B, C, . . ., in a solvent substance A, is a number defined in terms of the absolute activity λ_B by

$$a_B = (\lambda_B/m^\ominus)/(\lambda_B/m_B)^\infty = m_B\gamma_B/m^\ominus \qquad (T, p \text{ const.})$$

or in terms of the chemical potential μ_B by

$$\begin{aligned} RT \ln a_B &= \mu_B - RT \ln m^\ominus - (\mu_B - RT \ln m_B)^\infty \\ &= RT \ln(m_B\gamma_B/m^\ominus) \end{aligned}$$

where $m^\ominus$ is a standard value of molality (usually chosen to be 1 mol kg^{-1}) and where the other symbols are as defined in Section A.I.11.

It follows from this definition of a_B (compare Section A.I.8) that

$$(a_B m^\ominus/m_B)^\infty = 1 \qquad (T, p \text{ const.})$$

The name activity is often used instead of the name relative activity for this quantity.

The name relative activity with the symbol $a_{c,B}$ may be used for the quantity similarly defined but with concentration c_B (see Section 2.3) in place of molality m_B, and a standard value $c^\ominus$ of concentration (usually chosen to be 1 mol dm^{-3}) in place of the standard value $m^\ominus$ of molality.

Another relative activity, called the *rational relative activity* of the solute substance B and denoted by $a_{x,B}$, is sometimes used. It is defined in terms of the absolute activity λ_B by

$$a_{x,B} = \lambda_B/(\lambda_B/x_B)^\infty = x_B f_{x,B} \qquad (T, p \text{ const.})$$

or in terms of the chemical potential μ_B by

$$\begin{aligned} RT \ln a_{x,B} &= \mu_B - (\mu_B - RT \ln x_B)^\infty \\ &= RT \ln(x_B f_{x,B}) \end{aligned} \qquad (T, p \text{ const.})$$

where x_B is the mole fraction of the substance B in the solution. The rational relative activity $a_{x,B}$ is related to the (practical) relative activity a_B by the formula:

$$a_{x,B} = a_B m^\ominus M_A$$

A.I.13 *Osmotic coefficient of the solvent substance in a solution*

The osmotic coefficient ϕ of the *solvent* substance A in a solution (especially in a dilute liquid solution) containing molalities m_B, m_C, . . ., of solute substances B, C, . . ., is a number defined in terms of the absolute activity λ_A of the solvent substance A in the solution by

$$\phi = (M_A \Sigma_i m_i)^{-1} \ln(\lambda_A^*/\lambda_A)$$

where λ_A^* is the absolute activity of the pure solvent substance A at the same temperature and pressure, and M_A is the molar mass of the solvent substance A, or in terms of the chemical potential μ_A^* by

$$\phi = (\mu_A^* - \mu_A)/RTM_A\Sigma_i m_i$$

where μ_A^* is the chemical potential of the pure solvent substance A at the same temperature and pressure.

For an *ideal dilute solution* as defined in Section A.I.11 or A.I.12 it can be shown that $\phi = 1$.

Another osmotic coefficient, called the *rational osmotic coefficient* of the solvent substance A and denoted by ϕ_x, is sometimes used. It is defined in terms of the absolute activity λ_A by

$$\phi_x = \ln(\lambda_A/\lambda_A^*)/\ln x_A = \ln(\lambda_A/\lambda_A^*)/\ln(1 - \Sigma_i x_i)$$

or in terms of the chemical potential μ_A by

$$\phi_x = (\mu_A - \mu_A^*)/RT \ln x_A = (\mu_A - \mu_A^*)/RT \ln(1 - \Sigma_i x_i)$$

where x_A is the mole fraction of the solvent substance A in the solution. The rational osmotic coefficient ϕ_x is related to the (practical) osmotic coefficient ϕ by the formula:

$$\phi_x = \phi M_A \Sigma_i m_i/\ln(1 + M_A \Sigma_i m_i) = -\phi M_A \Sigma_i m_i/\ln(1 - \Sigma_i x_i)$$

A.I.14 *Relative activity of the solvent substance in a solution*

The relative activity a_A of the *solvent* substance A in a solution (especially in a dilute liquid solution) containing molalities m_B, m_C, . . ., or mole fractions x_B, x_C, . . ., of solute substances B, C, . . ., is a number defined in terms of the absolute activity λ_A of the solvent substance A in the solution by

$$a_A = \lambda_A/\lambda_A^* = \exp(-\phi M_A \Sigma_i m_i) = (1 - \Sigma_i x_i)^{\phi_x}$$

or in terms of the chemical potential μ_A by

$$RT \ln a_A = \mu_A - \mu_A^* = -RT\phi M_A \Sigma_i m_i = \phi_x RT \ln(1 - \Sigma_i x_i)$$

where the other quantities are as defined in Section A.I.13.

Note: The definition in this Section of the relative activity of the *solvent* in a *solution*, is identical with the definition in Section A.I.8 of the relative activity of any substance in a *mixture*. See also Section A.I.12.